ALLAN ARNOLD

I0474239

HOW WE PERCEIVE REALITY

OUR BRAIN DEFINES REALITY

outskirts press

PREFACE

This book was written to reconcile in my own mind the apparent difference between how we see the macro world we live in, which appears solid and substantial, and the micro world of atoms and particles, which does not appear to be solid or substantial at all.

The field of quantum mechanics and the atomic structure of matter suggests that the micro world is almost empty space with no substance at all; there is almost nothing there--just very small particles of matter with large distances between them. Yet this is what we are made of. What gives my brain the illusion of a solid macro world when in reality there is almost nothing there? How do we perceive the macro world around us?

From the earliest times of man's existence, he has wondered how the earth and its many life forms were created and why he is here. Various myths and stories have been proposed regarding how the earth was formed and also how man came into existence. These myths and stories do not stand up to scientific scrutiny or in many cases common sense. These stories did, however, satisfy man's hunger to understand how he came into being and the existence of the world around him. This was usually attributed to a superior entity beyond his comprehension. Throughout written history he was also puzzled about how he perceived reality-- what was the mechanism in himself that enabled him to see or feel? Why did he have abstract thoughts?

From the beginning, man has used his intellect to develop tools for his everyday use and to harness nature for his purpose. This attribute has led over time to the development of powerful instruments, which he has used to examine both the macro and micro world. By using these instruments and his intellect, he has studied the universe and found our home, the earth, to be a small planet circling an ordinary star in a galaxy consisting of billions of stars, in a universe with billions of galaxies. He has found from careful observation that the universe he lives in is expanding; from this and other observations he has theoretically deduced how this Universe was created many years ago in a cosmic explosion called "The Big Bang."

At the moment of creation, the universe was very small and consisted of a massive amount of pure energy at a very high temperature. This energy field began to expand immediately after creation and continues to this day. As it expanded, it cooled; some of the energy was converting into mass, and

ALLAN ARNOLD

minute particles of matter appeared in the energy field with positive and negative charges these coalesced in to the primary element of the universe hydrogen and created a cloud of pure hydrogen, the beginning of a material universe. This cloud of hydrogen was not spread uniformly in the proto universe, but varied with areas of different size and density; gravity, one of the primary forces of nature, collapsed these various and different-sized areas into the billions of galaxies that make up the universe we observe today. Within each galaxy, stars were born, using the same mechanism of gravitational collapse that formed the galaxies. This process can still be seen occurring today

By studying the biological structure of genes and DNA within us and other life forms, and the development of other creatures as they adapted to their environment, scientists theorized how we evolved from simple life forms to the complex creatures we are today. Nature has, by using the process of evolution over eons of time, created the most complex structure we know of: the human brain. With this structure, man has made it possible for him to interpret the world we live in and make him conscious of his being. Without the brain, there is no existence on our planet--only man, with his large brain, knows that we exist on a small planet in a vast universe.

At the beginning of the 20th century, scientists discovered the theory of General Relativity, which explained time and gravity in a more accurate way than Newtonian physics, and also a new field of science called quantum physics, the study of the micro world of atoms and energy fields. This new field of science has its own set of rules and laws, which appear very different from the Newtonian laws of physics we have used for years to describe the motion and substance of our world. We are part of this field of science; everything in the observable world and universe, including us, is made of atoms and energy and therefore must conform to the rules and laws of quantum mechanics. We are just beginning to understand the impact and possibilities of this new field of science.

The planet Earth consists of a macro world in which our life form exists--where we live and interact with the environment and each other. This world is all around us. Our senses give us the impression that it is real. We live our lives with the conviction that it is real, but we know the earth and its many life forms are made of atoms and energy which our senses cannot discern. There is whole micro world out there that we do not perceive with our limited set of senses, but nevertheless, it is still there.

Humans have developed instruments and theories to analyze this micro world and found it to be a strange environment of particles and energy fields that behave counter to what we consider logical thinking, but which govern all aspects of the macro world we live in and which our senses can discern.

Our Earth and everything in it was a product of nature until humans arrived. Almost all of the natural world evolved without human intelligence--some animals created structures, such as bird nests and beaver dams, which had little impact on nature.

With the appearance of humans on the Earth, a completely new paradigm was initiated. "Nature" no longer was the only source of innovation in the material world. "Nature" had evolved its single most complex structure, the human brain. The complex brain had given humans the capacity for abstract thought and language, two powerful attributes that have made it possible for humans to be aware that we are part of a universe so immense it is almost beyond our comprehension and become conscious of the individual self, living on a small planet circling an ordinary sun in the Milky Way galaxy, which is just another galaxy in the immense universe.

One of man's attributes is his imagination! He has built structures and machines and developed agriculture to feed himself; he has made dams, diverted rivers, and organized cities and towns and altered the face of the Earth in many ways for his own purpose. The cities concentrate the brain power of humans, speeding up the rate of technical progress through mutual interaction between brains and facilitated by the use of language.

By studying the cosmos, scientists with their knowledge of physics have theorized how stars were formed by gravity collapsing clouds of hydrogen gas. As the cloud collapsed, the pressure and temperature in the collapsing hydrogen cloud became enough for nuclear fusion to begin, and the star began to shine. There are several types of stars, from super giants to dwarf; our star is a main sequence star of average size. As far as we know, this is the basis of star formation, sometimes this process of star formation would create a much larger star--these stars had short life spans, used up their nuclear fuel quickly, and became unstable, eventually exploding in a supernova explosion. The heat and pressure of the supernova explosion was enough to create all the rest of the heavier elements of the periodic table.

The force of this explosion scattered these atomic elements into space, where they mixed with the hydrogen clouds from the (BB). As newer stars were formed in these mixed clouds of hydrogen and other atomic elements, the gravity of a newly forming star attracted this debris from the supernova into the orbit of a new star. Gravity gradually caused this debris circling the new star to coalesce into rocky planets like the Earth. The atoms in our bodies, the planets and all things on the planet came from this debris. The atoms in our bodies and the planets came from exploding stars, which were originally just energy. The energy came from the (BB), which came from nothing. This is hard for us humans to comprehend--we cannot imagine something from nothing. Our brains are not capable of understanding this concept. Einstein's equation $E=mc^2$ shows the relationship between mass and energy. Since (c) is the speed of light, a very small mass creates an enormous amount of energy. An atomic explosion is an example of converting a small mass into a huge amount of energy. Much of the early cosmos is locked up as mass so we, the Earth, Sun etc. are just another form of energy! We will show how energy and the other primary forces in nature are involved with our senses and brain in perceiving the macro world.

There are approximately seven billion humans living on the planet Earth. They all have large brains, have five senses, and share a similar view of their surroundings. They all have the ability to see, hear, and touch with the same set of senses;

ALLAN ARNOLD

because of this, every person's perception of reality is about the same. We all see trees, rocks, people, and everything else the same. If one of our present senses was changed, or another sense added to all humans at the same time, our perception of reality for everyone would change, but no one would notice any difference--everyone would be seeing the same way. If one of the senses—sight, for example--was replaced by another method of seeing, our perception of reality would be entirely different, but nonetheless just as real to the humans with the altered set of senses. Imagine using X-rays to see rather than light--the view of reality would be the same for everyone, and no one would realize they were seeing with X-ray radiation. No one would doubt it was the real world they were seeing; the brain would have no concept of light, just as it has no concept of seeing with X-rays. There would be no color, just shades of grey. How the brain would see objects illuminated with X-ray radiation is open to conjecture. Our present sense of sight converts photons of reflected light from the objects we perceive in the world around us. The photons enter the eye through a lens which forms an image on the retina. The retina consists of light-sensitive rods, which convert the photon energy into electrical energy, which is transmitted to the brain. The brain interprets what we see as the real world around us as a mass of electrical signals forming an image in the brain.

This mechanism for observing our world as evolved over time using some form of energy; it uses the energy in a photon and converts it into electrical energy. How it would have evolved using X-rays is unknown--a life form could evolve where sight is replaced by an entirely different sense. We know our eyes and brain have been evolving over billions of years of evolution, adapting to a narrow band of radiation called light. This can be split into various colors by the action of white light reacting with the atomic structure of the atoms in the macro world. Why the sense of sight developed in this way is open to conjecture; there are many other radiation bands which could have been selected. Our sun has a color temperature of 5000K, so white light of this temperature is illuminating the Earth; life forms evolving on our planet would naturally develop the sense of sight using the sun's color temperature.

In the early history of the Earth, only primitive life forms existed; they had no brains and none of the five senses we have. These primitive life forms were only vaguely aware of their surroundings. They could not see or hear; the Sun and Earth do not exist for these primitive life forms. To see the macro world, a life form must have a full complement of senses and a brain to observe the world we live in.

Our reality is the world we live in. It is the macro world we observe and interact with. Heisenberg, one of the early scientists working in the field of quantum mechanics said the only reality is what we observe--"There is no other reality."

The average human has a body, a brain, and a complement of five senses that allows us to live and interact with macro world around us. These attributes are needed to make sense of the massive amount of information constantly fed into our brain. The macro world is made from the atoms of the micro world. The atoms in the micro world are made from particles, protons, neutrons, electrons, and the electromagnetic energy between the particles. These particles make up the structure

ALLAN ARNOLD

of the atoms and molecules, forming the substance we see and sense as the macro world.

Our senses are activated by the interaction of the micro world's radiation energy, gravity, and the atomic forces in the atoms and molecules that make up the substance of the macro world. Our senses react in various ways to the four forces of nature: gravity, electromagnetism, the atomic weak force, and the atomic strong force. The interaction of our senses and these four forces of nature give our brains the appearance of what we observe as reality. We will discuss how our senses and these forces combine to create our reality.

The essential item that links our senses and the forces of nature together is the human brain. Without the brain, we would be no better than primitive microbes-- no music, no brilliant sunsets, no sensation of love and happiness, not even awareness of our existence.

The brain is a multicellular structure made from the atoms of the micro world into molecules and cells required for the brain to function. The brain uses electromagnetic energy to communicate between different elements of itself and the nervous system of the body. The brain observes the world around it by storing and interpreting the huge volume of information coming through its five senses into its memory and at the same time allowing its life support system (the body) to sustain its brain and continue to react to the forces of nature. The brain is always storing the information it receives and from this knowledge builds its own reality. It is constantly learning about the world it observes and lives in.

The brain and body develop from a single cell egg and a sperm into a multicellular organism. The brain is part of this organism and consists of trillions of cells, each adapted to a special purpose. It is the most complex structure developed by the natural world. The brain learns by input from the forces of nature. Light radiation allows the brain with its eyes to see objects in its environment. The brain also learns from teaching by other humans. For instance, a child will learn what a tree looks like by observing the tree through its sense of sight, and it will associate the word "tree" with that object because other humans can also see the tree and have a word for it. That particular word is stored in the brain and is enunciated by electrical signals generated in a special section of the brain and sent to the body's voice mechanism, which creates pressure waves in the air; the pressure waves travel through the air to the ear, where another mechanism converts the pressure wave in the air to an electrical signal, which is sent to the receiving brain, where it is recognized as the appropriate word for the object being viewed namely tree.

This process is called language and is a powerful method for brains to communicate with each other. The brain has also learned to communicate using the sense of touch or sight—Braille, using the sense of touch, or sign language and writing, using the sense of sight. When reading what amounts to marks on paper can become for the reader an emotional experience of a story or to a scientist a complex mathematical theory describing a natural phenomenon or idea, these marks on paper can be arranged to represent almost every thought or idea in man's knowledge. In reality, the brain creates these thoughts within itself. The words represented by marks on the paper convey

ALLAN ARNOLD

the thoughts of another human, "the writer," into the reader's brain. Those thoughts can be a story, news, information and so on.

As the brain develops and matures, it also acquires cognitive ability to perform difficult tasks such as walking, balancing on a bicycle, solving problems, etc. The brain constantly increases its knowledge about the macro world it lives in, and at some point in its development, the brain becomes conscious of itself and other humans living in the same environment.

The human race is the result of evolution that began as primitive life forms billions of years ago, in a world very different from what we live in today. These early life forms developed at first in the sea and slowly migrated onto the land, where they were instrumental in changing the earth's atmosphere to oxygen-rich. These early life forms had no brains and reacted only to the local environmental conditions; they gradually acquired senses and reacted to changes in their surroundings and became more aware of their environment. Shellfish such as clams, living in the sea would open to feed with the incoming tide. They have no brains, but they have a sensory and nervous system to open their shell at the appropriate time to feed.

A fish has a primitive brain, but it is not aware it is swimming in the sea; the fish does not know anything different. It is aware of its surroundings because it has primitive eyesight--it needs this sense for survival. Without sight, it would be unable to escape a predator or swim in shoals for mutual protection. These senses were often acquired by evolutionary pressure for survival. Our own eyes can see with far greater precision than a primitive fish eye. We can focus our eyes for a sharper image, see color, and judge distance with our two eyes. To utilize these capabilities, the brain increased in size to accommodate these new attributes. As the brain developed in various life forms, it utilized the increased capacity and cognitive ability to more effectively use these senses, until in humans it has improved to where we are aware, not only of our environment, but conscious of our existence, which leads us to ask the question "What is the real world?"

With their large and complex brains, humans have, with their cognitive ability, memory, and observation developed the arts and sciences and organized societies into cities and states. They have used the brain to discover the underlying nature of the macro world we live in (Earth) and our place in the Milky Way Galaxy and the Universe. One of the sciences investigated is the micro world of atoms and the field of quantum mechanics. This has shown the micro world to be almost empty space. To give some idea that what we perceive as solid, such as a rock, is actually not solid at all, consider this: take one of the atoms in the rock--if the nucleus of the atom was the size of a football and was placed at the center of a football field, then electrons circling the nucleus would be the size of a pea on the goal line. There is plenty of empty space in the atoms that make up the rock, which to our senses appears solid.

Obviously, what we perceive as solid is patently not so. Our senses are reacting to something else to give us the apparent illusion of substance and solidity. Most radiation energy simply passes through what we perceive as solid. Our bodies are constantly exposed to all kinds of micro wave radiation, yet we

ALLAN ARNOLD

do not feel or sense this radiation; it passes through us without interacting with the atoms in our bodies. Some particles, such as the neutrino, can pass completely through the Earth without interacting or striking another atomic particle. An exception to this is light and radiation close to the band of frequencies that our eyes have evolved to see, which is fortunate, because light interacts with most of the atoms making up solid structures. Light has other attributes that enable us to have the illusion of color. Even light can pass through many gases without interacting, and also some solid structures. Glass is transparent to light but is made from substances opaque to light.

We view the world around us with our senses. They consist of three primaries: sight, hearing, and touch; and two secondarys, taste and scent. These senses react to the forces of nature to give us our concept of reality. Our senses are tuned to the forces of nature, gravity, the electromagnetic force, the atomic weak force and the atomic strong force--the last two act only indirectly on our senses. Our senses are tuned to react to these fundamental forces of nature through the long process of evolution, taking billions of years of trial and error and variations in our DNA code to accomplish. The forces of nature have always existed on the planet Earth, and nature has evolved our senses to react to these forces of nature. The human brain has developed a reality based on our senses and the forces of nature. We can all feel the effect of gravity on our bodies--even animals can feel it, although they may not realize it. We know that without gravity, we would float around like an astronaut in space. We can see because of the electrical charge in the atoms that make up every element in our world. When looking at an object we perceive as white, it is because the light radiating from the sun is made up of photons vibrating at different energy levels over a narrow band of the electromagnetic spectrum we call light. When all the different energy levels are mixed together and reflected from a surface or object, it appears as white to the eye. In this spectrum we call white light, there are photons with different frequencies and energy levels; these differences correspond to the colors we see in the objects around us. Blue is more energetic than red, and because of this disparity, we have color vision.

The eye is connected directly to the brain and gives us the sense of sight. Photons of light enter the eye through a lens and are focused onto a system of light- sensitive nerves called rods. When photons from a light source such as the sun strike the surface of an object, the electrons in the atoms or molecules on the surface of the object react by dropping to a new orbit and emitting a photon. If the energy level of the emitted photon is at the red end of the white light spectrum, the rods in eye perceive it as red--not because it is red photon but by its energy level. This energy level sends an electrical impulse corresponding to red down the optic nerve to the brain, which the brain interprets as red; color is a construct of the brain. We know that this is red not because of some intrinsic value in the object being observed, but because but we have learned that electrical energy of this magnitude as red. We could have called it anything we liked, but in the English language, it is red. The brain has the ability to differentiate colors even under varying values of white light, both sunlight and incandescent; a photograph taken with daylight film under incandescent lighting will have a yellow cast, yet the brain does not see the yellow cast; it views the colors as normal. Color is a construct of the brain and not some intrinsic attribute of an object being viewed.

The sense of touch gives our brain the appearance of solidity and substance to everything in our world. When we pick up an object, we feel its weight (gravity) and our sense of touch tells our brain it is solid. What you feel when your finger touches the surface of an object is the negatively charged electrons in your finger being repelled by the negative charge in the electrons on the surface of the object. Our nervous system sense this and sends an electrical impulse to the brain to give us the sense of touch. What your brain senses as solid object is the electromagnetic force between electrons. The object is not solid. What you feel as solid is just the interaction between the electrons in your finger and those in the object--your finger actually never touches anything. In reality, almost nothing is there. If you touch something hot, the atoms in the hot object are agitated compared to the atoms in your finger; this causes the atoms in your finger to agitate more vigorously. Your nervous system senses this and sends an electrical signal to the brain. The hotter the object, the more violently the atoms in your finger vibrate; the stronger the vibration, the more powerful the electrical signal being sent to the brain, and the cognitive ability of our brain will tell us to pull away from the hot object. Solidity is a construct of the brain, using the two senses sight and touch. The brain acts only as a device to interpret the macro world around us. Our consciousness tells us what the brain perceives.

We can feel the force of gravity and see the effect it has on objects; when an object is dropped, it falls and stops when it reaches the ground. This is caused by the force of gravity. It does not pass through, because the force of gravity is opposed by the negative charge in the electrons in the falling object being repelled by the negative charge in the electrons in the ground material. If the object is dropped into a liquid such as water, the negative electrons in the object and water are still there and offer resistance to the falling object, but the atoms and molecules are not tightly bound together as in a solid and can slide around against each other so the water just moves out of the way, allowing the object to pass through.

Hearing relies on pressure waves in the mixture of gases all around us we call air. A sound source, which can be human voice, a bell, or any other noise-making item generates a pressure wave by agitating the air molecules in contact with the sound source. These molecules agitate the molecules next to them, and so on, forming a pressure wave radiating out from the sound source. The wave travels through the air at the speed of sound. When the pressure wave reaches the ear, it vibrates your ear drum; the vibrating ear drum generates an electrical signal in the auditory nerve which is sent to the brain. The electrical signal from the auditory nerve is different; the brain can differentiate the electrical signals from each of its senses and perceives the signal from the ear as sound. We can look at a scenic view, feel the wind on our face, hear the birds sing, smell the flowers, and taste the wine all at the same time. All of these electrical signals are sent to the brain together at any instant; our perception of reality at any instant is a mass of electrical signals sent to the brain by our senses. It is our consciousness which perceives the difference in the signals, sorts them out, and gives us our impression of reality.

ALLAN ARNOLD

From the forgoing, it is apparent we perceive our world as a mass of electrical signals sent to the brain by our senses, which are activated by one or more of the fundamental forces of nature. Without these forces of nature and our senses, we would perceive nothing. When we look at nature without these senses, we find that the micro world is almost empty space, with almost nothing there. An atom is made up of particles, called protons, neutrons, and electrons; these are minute in size relative to the size of the atom. Each of these particles has different properties and plays a significant part in allowing our senses to perceive what we call reality.

What we see or feel as solid matter or a definite object is a result of the fundamental forces of nature reacting with our senses. What we are seeing or touching is not solid at all; it is mostly empty space. Our senses give us the impression that there is something there, but in reality, it is an illusion--a construct of the brain.

One of our primary senses is sight. This relies on a narrow band of frequencies In the electromagnetic spectrum, ranging from gamma rays with wavelengths measured in microns to radio waves measured in meters. The band of frequencies our eyes are sensitive to is called white light; it has the unique property being split into three primary colors: red, green, and blue. Our eyes and brain have evolved with the ability to distinguish these colors and various combinations of them to give us color vision: when we perceive an object, our brain, with the sense of sight, tells us it has a given shape, size, and color which we can identify from our memory. If our eyes had evolved sensitive to other frequencies in the electromagnetic spectrum, our perception of reality could be very different. It is difficult for us to imagine how our view of reality would be if our eyes were sensitive to either the X-ray or infra-red band of frequencies.

We are insensitive to much of the electromagnetic spectrum which permeates our environment; for instance radio waves are all around us, but we are totally unaware of them. Switch on a radio or TV, and you will see they are there, because we have invented devices designed to decipher the radio waves surrounding us and convert them into sound or visual images that you can interpret with your senses, if you switch off the device propagating the radio wave you will get nothing and you will not have known whether it was on or off except for the radio or TV. The Earth and the Universe are full of electromagnetic radiation; some of it is harmful, and to most of it we are totally insensitive.

Sight is the ability of our brain to visualize objects. Humans have two eyes connected to the brain for this purpose. Nature has evolved the eye to be sensitive to atomic particles (waves) called photons so that the brain can visualize objects, and identify them with objects previously stored in its memory. If we are in a room with an energy source generating photons (light), we are in what is called a photon field. These photons are surrounding us, bouncing off all the walls and objects in the room; switch off the light source in a blacked-out room, and there are no photons; we have lost the sense of sight. The objects are still there, but to our brain, they only exist in our memory. While the lights are off, change everything in the room instantaneously, then switch them back on--we are in a new reality. What was in the room previously is now memory and in the past; what

ALLAN ARNOLD

will be in the future has not happened. From this thought experiment, we can see that REALITY is what we observe at any instant of time. Everything else is in the past or future.

Our visual perception of reality relies on the radiation called light. If it were otherwise, our view of the macro world would be very confusing. For instance, a leaf in sunlight appears to be green, but in infra red radiation it has no color. So our brain is what tells us the color of the leaf; the eye is sensitive to the green frequency in white light and sends an electrical signal to the brain, which interprets it as green. The photons of each color have a different frequency which the eye detects and converts the photons of a specific color to an electrical voltage for each color. The brain interprets these differing electrical signals as color. A child, when it is very young, can see in color--we know this from the structure of the child's eye. It does not realize it is seeing in color; as it grows older, it learns to associate various colors with names and can visualize colors--the concept of color is learned. Our ability to discern color gradually disappears as the light intensity gets dimmer. In moonlight everything appears as shades of grey.

Light radiation is transmitted as waves or particles called photons. When a substance is heated to a certain temperature (about 5600F) it radiates photons in our visual spectrum. The Sun is emitting copious amounts of radiation in many different frequencies; all are illuminating our planet Earth. The visible light spectrum is made up of radiation of different wave lengths and energy levels. This is important for our sense of sight, because our brain uses this difference in energy and frequency to define color.

The Earth has ninety-six elements in its makeup. All of these elements—solid, liquid, or gas--are found either as separate entities or in combination with other elements to form the almost infinite variety of substances found on Earth.

A leaf from a tree is a common object that consists of a mass of elemental atoms mixed together to form molecules in the shape of a leaf. Our brain has learned that a mass of molecules bound together in this way is probably a leaf. We see the leaf because it is illuminated by white light from the Sun. The leaf contains the chemical chlorophyll, which reflects a frequency in the white light spectrum we perceive as green. The brain recognizes the leaf from memory and the green color from the photons emitted by the molecules in the leaf. The perception of the leaf in the brain is a solid object with color, but how the brain connects the two is not easy to explain; the leaf could be yellow in autumn, and the brain would know this. If we examine the leaf in the micro world of atoms, we find that there is almost nothing solid and substantial. What is there is empty space with particles of matter in prearranged patterns we call atoms, combined with other atoms forming molecules with large areas of empty space between the particles. The brain has, with its senses and the forces of nature, perceived an object that is solid and substantial when in reality there is almost nothing there. The leaf is a construct of the brain using the forces of nature and the brain itself with its complement of senses to provide the necessary data for the brain to recognize and identify an object.

In a previous chapter, we explained how our senses built a picture of reality in the brain, using the fundamental forces of nature and our senses. The brain is a product of nature and is made of atoms and molecules of the micro world. The atomic strong force holds the nuclei of the atoms together to form the stable elements necessary to form the molecules and chemical compounds that make up the structure of the brain. Without this force, there would be no elements to form the complex molecules found in the brain, or any other structures. The weak force binds electrons to the atomic nuclei and plays a role in our ability to see.

The adult brain is the product of our genes and millions of years of evolution and many changes to arrive at its present state. The brain begins developing in early life when the mother's egg is fertilized. It has no knowledge or intuitive ability; it is just a mass of cells growing and dividing according to the rules of nature to form an organ, the most complex in nature--the brain. While it is growing inside the mother's body, the brain is adding trillions of cells and connecting to the rest of the life form: the heart, lungs, nervous system, etc. These body parts are necessary to have a viable life form able to survive and function outside the mother's body and in the Earth's environment. Apart from some intuitive capability, the brain has very little knowledge and acquires only limited information through its senses about the outside world while inside the mother's womb. The developing brain knows almost nothing about the macro world it will be born into. The brain is essentially empty of any knowledge but has the capacity to absorb massive amounts of information. Part of the brain has a subconscious component to control the functioning of the organism necessary to keep the brain alive once it has been born into the macro world outside of the mother's body. At the beginning of life several preprogrammed events take place, such as breathing, or are learned immediately, like its mother's face--this information is stored in the brain. The brain continuously stores information coming through its senses from the macro world in which it now lives. The brain keeps on storing information as it grows, building a picture of what it observes as reality, which is nothing more than what its senses transmit from the macro world around it. This world is made of atoms and energy, nothing more, which our instruments tell us is mostly empty space.

Take the leaf again. Our senses perceive it as something real and substantial we can pick it up, look at, and handle--our senses of touch and sight convince our brain that it is real. What our brain is telling us is the reflected image of green light photons (energy) from the chlorophyll molecule, which is part of the atomic structure of the leaf, is what our brain has learned to be a leaf. If all the light spectrum were reflected from the leaf, it would appear white; if the leaf is viewed in the fall, chemical changes in the leaf makes it look red. It is still the same leaf, only now red photons are reflected from the atomic structure of the leaf instead of green.

The leaf is actually a mass of atoms and molecules shaped like a leaf. If it did not reflect any light we could not see it when it is dark there are obviously no photons, therefore we cannot see it, but we can tell something is there because if we try to touch it the electrons in the leaf will be repelled by the electrons in our fingers giving our brain the illusion of solidity. In the micro world the leaf is not solid; it is a mass of particles of minute size with empty space between them. The actual mass of the particles is minute compared to volume of the atom. If all the particles in one cubic

ALLAN ARNOLD

centimeter of rock were crushed together with no spaces between them it would have a mass equivalent to Mount Everest.

What we observe with our sense of sight are the green or red photons reflected from a leaf illuminated by a white light source such as the Sun. It takes the brain to identify the object as a leaf, because this image of the leaf has been imprinted on our brain's memory and learned from experience that it is indeed a leaf. If we look at an object we have never seen before, we can still see the object but not identify it; we can make a guess if it is something similar to what we have in our memory. If it is not in our memory, then we are completely mystified. A person from Africa seeing snow for the first time has no idea what it is. The eye is an optical device that can see an object only if there are photons reflected by the object reaching the eye. The eye is connected via the optic nerve to the brain. It is the brain that visualizes and identifies the object being viewed from information stored in memory. This attribute of the brain can be dynamic, like playing a game or sport where the brain interacts with its own body, or static, like reading, where it interacts only with its memory. Our brain can only identify an object if there are photons reflected by the object for the eyes to observe, or sound waves for us to hear, or objects for us to touch. An example where all of these senses are absent can be experienced in a dark anechoic chamber. The only force you can feel subconsciously is gravity. You cannot see, feel, or hear. There are no light photons, sound waves, or items to touch. Your senses are deprived of any input; it feels uncomfortable and out of this world. You have the feeling that nothing exists. You open and close your eyes as the brain tries to obtain some sensory input. If a light is turned on the feeling goes away immediately with the sensory input of sight. Even without sensory input the brain does not switch off and go blank; the macro world remains in the brain and is its reality.

A baby's brain, once the child is born, starts to build its own reality, immediately. It begins to acquire knowledge and cognitive ability necessary to sustain life. At the same time it is acquiring cognitive ability; the brain is also absorbing information about the macro world provided by input through its senses and the forces of nature. The brain does not observe the micro world of atoms and molecules--only the interaction between the forces of nature and its senses. To man or animals the micro world of atoms and molecules does not exist; you cannot see or touch an atom or molecule, but you can visualize or touch a mass of atoms and molecules in the form of an observable shape such as a leaf. The brain cannot see or feel what everything is actually made from, only the interaction between the forces of nature and its senses. However, by using his cognitive ability and observing phenomena occurring in the macro world such as magnetism and electricity, he has discovered the existence of the micro world of atoms. Using instruments, mathematics, and his cognitive ability he has explored the atomic structure of matter and found it to be strange world of minute particles and energy behaving in an unpredictable way.

With powerful instruments, his intellect, and his knowledge of physics he has discovered how his senses and the forces of nature interact to form an image in the brain he perceives as reality, which is solid and substantial. His instruments and studies show a different story. What one sees and feels as solid and substantial is actually almost empty space--there is nothing there! The atoms making up the macro world are empty of solid matter. The perception of reality is nothing more than electrical signals imprinted on the brain by the interaction of the brain's senses with the forces of nature. The question arises: What is out there to form our reality? Just particles of minute size compared to the volume of each atom contained in our observable universe. Each atom is essentially a self-contained unit with protons, neutrons, positrons, etc., bound together by the forces of nature and energy fields. The Standard model of particle physics describes all the particles found so far, the latest being Higgs-Boson. There are ninety-six different atoms from the simplest hydrogen, to the most complex, plutonium. These atoms are what we are made from.

If our senses or the forces of nature were any different from what they are now our perception of what we observe as reality would not be the same. The study of physics tells us the forces of nature are the same throughout the observable Universe then the only variables in our perception of reality are the brain and its senses. Without the brain there would be no reality--the moon doesn't know it causes the oceans on earth to have tides; it has no brain to perceive this reality, but brains on earth know this to be true from observation. Without the brain there would be no reality. "Things" would exist, but unless there are brains to observe them the "Things" are not part of our reality.

Animals on the planet earth build structures for a reason, whether it's a bird's nest or a 747 jet airliner, there is always a need or reason for designing or building the item. Nature has by a process of evolution, trial and error, and over a long period of time built a complex structure, the human brain, for what reason? no one seems to know-- Unlike the 747 airliner, which is designed using logical engineering and processes to arrive at its final configuration. On the other hand the brain seems to have arrived at its present configuration by trial and error and chance; there seems to have no coherent attempt to design the brain by a process of planning and logical design. This large brain has evolved for us to become conscious of our surroundings and also observe what our brain thinks as reality.

There can be other realities beyond our comprehension; on the planet Jupiter, nothing we perceive as real would exist. Life forms on Jupiter, if they existed, would have an entirely different view of reality. Jupiter consists of almost all gas, so intelligent life in that environment would be very different. Can nature evolve a life form suitable for Jupiter? All we know is that nature evolved our intelligent life on the planet Earth over a long period of time. If brains existed on Jupiter, they could not envision our macro world, but they would have their own reality. In fact any world in the universe with a different set of senses would perceive their own reality depending on the environment on that world and how their brains reacted to the forces of nature. Our life form is the result of evolution over a very long period of time on a planet that supported our life form's growth and development.

Many of the items created by man reinforce his concept of reality. They are the result of his imagination, an attribute of the brain. Every time humans construct an item that is not part of the natural world, they think that they have created something real. A jet liner aircraft would be a totally alien concept to a 176th-century person; the thought of travelling in a metal tube at 30000 ft in complete comfort, with the air passing by at 600 mph just a few inches from his head would be incomprehensible and not part of that person's reality. Yet, it is part of ours. The concept of the jet plane did not spring into existence out of thin air. The plane started in the brains of humans, then using their cognitive skills, plans were made, parts were fabricated and assembled to form the jet plane. Each person working on the various parts of the plane saw the item he was working on as real and substantial when in reality it was not. The end result is a plane consisting of millions of parts made by man. It is, however, still a collection of atoms assembled together obeying the forces of nature with a useful purpose for humans.

The aircraft flying at high altitude is an example of man harnessing the micro world's primary forces to create a benign environment for humans to live where the world outside the aircraft is completely hostile. The only item protecting the benign environment containing the passengers is a thin shell made from aluminum atoms consisting of neutrons, protons, and electrons. This shell is transparent to many forms of radiation; there is plenty of empty space between the particles in the aluminum atoms for the radiation to pass through, but the negative charge on the aluminum atoms repels the negative charge on the air molecules, preventing the inside or outside air from passing through the aluminum shell, keeping the two environments separate. Nature has evolved structures to do the same--a prime example is a bird's egg. Nature surrounds a fragile embryo with a calcium shell for protection against the elements.

What is reality? Is it anything the human brain can think of? If so, where will it end, where will our imagination take us? We have progressed from the wheel to space flight and beyond; and still our imagination of things to come does not show signs of stopping. Where will man's insatiable thirst for technology and science end? His constant search to uncover the workings of the natural world starts in the brain and progresses by trying to find a formula or explanation for what has thought about. A prime example of the brain discovering something new is a phenomenon in our universe called a "black hole." A scientist serving as a soldier on the eastern front during WWI theorized this phenomenon in his brain. At first it was considered an outlandish idea--the idea of an object in space where gravity is so strong that nothing can escape from its gravitational field, even light, seemed beyond comprehension. We now know that black holes do exist. What new ideas will become part of our reality when we fully understand the strange behavior of the micro world of quantum physics? Will it be like the field of Newtonian physics, which opened a new realm of understanding of the natural world, or the consequences of deciphering the human genome? These are the physical results of man's imagination.

Nature has built a reality on this Earth base on the four forces of nature and his five senses. As far as we know the four forces of nature are the same throughout the Universe, but there could be other senses we are not aware of. Many birds have a strong sense of direction; a pigeon can unerringly find its way home using its senses in a way we cannot fully comprehend, it has been suggested that some birds, including the pigeon, use the Earth's magnetic field to migrate from their winter to summer home. We know from using the compass to navigate ourselves around the world that the Earth has a magnetic field,

ALLAN ARNOLD

we know what causes it, and have made an instrument, "the compass" to use it, but our senses cannot perceive this magnetic field in any way. In fact, there are many radiation fields we know about from observation and what our scientific instruments tell us. We cannot perceive many of these radiation fields with our senses; if we could our sense of reality might change dramatically.

We can trace our heritage all the way back to primitive life forms. Nature has evolved countless life forms over millions of years. Most have been discarded by catastrophic events, climate change, or become extinct by natural selection--why so many iterations to arrive at us? Man uses intelligent design to achieve an objective such as from primitive flying machine to a complex jet airliner in eighty years, an extremely brief time frame compared to nature--is this because man is intelligent and uses that intelligence to accomplish his objective. Orville and Wilbur only wanted to fly--they did not envision a jet plane; that came about through man's desire to travel to anyplace on the Earth rapidly or as a weapon of war. Nature does not work that way; it evolves a life form in an unpredictable and inefficient way called natural selection, over millennia of time. We and countless other life forms are the result of that method.

Humans have developed through many iterations a complex structure, the brain, we are the only species on this Earth to appreciate the two worlds the one we live in--the macro world and the one we are made from, the micro world.

The brain, with its five senses and its memory, gives us our concept of reality on this Earth. This is the apparent physical reality all of us perceive; there is no other physical reality that we can determine. We know there are other forces we cannot perceive with our set of five senses--magnetic fields, for example. There are also other natural forces and energy fields which could possibly change our perception of reality, if we had the senses to detect them. It is the interaction of our senses with the forces and energy fields of the micro world that give our brain the perception of the macro world.

Our intelligence tells us the macro world is real and not an illusion because everything we perceive consists of atoms arranged to form an object such as a leaf, a metal spoon, or the wind on our face--each object is made from atoms that are mostly empty space, which implies that there is nothing there and everything we perceive is an illusion in the brain, but it is still an object, a collection of atoms in space that we can perceive with our senses and recognize from our memory as an object, whether the atoms are formed into an object by nature or humans.

With our senses, the forces of nature, and our brain we perceive the macro world we live in; our brain also has the capacity for abstract thought. This ability can be highly developed in some humans--inventions, music, technology, and science all began as ideas in the brain. Only humans have the capability to convert ideas and abstract thought into items useful for our well-being, pleasure, and enlightenment; Mozart for his music and Newton for his science are typical examples of the human creative brain. This creative ability is the most important attribute of the brain; from this, humans have developed civilizations, traveled into space, and explored every corner of Earth.

Our brain is the most important organ in our body. It allows us, through our senses, to perceive the macro world created by nature; when it dies, there is nothing to interpret the forces of nature, which still remain and do not die. If the brain is dead, then to a human there is no macro world to perceive, so death becomes oblivion to a human living in the macro world. Many humans believe in a spiritual world which survives after the brain dies; there is no physical evidence for this--just conjecture and wishful thinking.

Although the human brain is the only way we perceive our macro world, there are forces and energy fields we do not perceive but know they exist. Nature, by a long process of evolution and trial and error, has developed the human brain for the environment found on the planet Earth and a select collection of natural forces embodied in the atomic structure of our makeup. To say that humans are the only way nature could have developed a big brain for intelligent life, given the huge variety of life forms and climatic conditions found in nature, is not a reasonable conclusion.

Until humans arrived on the planet Earth, nature, with all its varied life forms evolved and developed in a haphazard way guided by a process of natural selection. Humans possess intelligence and have created civilizations and technology that changed the Earth dramatically over a short period of time. Using human intelligence could it be possible to change the way we evolve in the future.

Everything we perceive is a collection of atoms floating around in space, shaped into objects our brain perceive as real--but in reality, these collections of atoms are mostly empty space.

www.ingramcontent.com/pod-product-compliance
Lightning Source LLC
Chambersburg PA
CBHW072030190526
45166CB00015B/1702